The Wonders Within: Exploring the Human Body

Introduction:

The human body is an incredible machine, full of wonders and mysteries that continue to fascinate us to this day. From the smallest cells to the largest organs, our bodies are constantly working to keep us alive and healthy.

In this eBook, titled "The Wonders Within: Exploring the Human Body", we will take a journey through the amazing intricacies of the human body. We will explore the different systems that make up the body, such as the circulatory system, nervous system, and respiratory system, and how they work together to keep us functioning.

We will delve into the structure and function of various organs, including the heart, lungs, brain, and kidneys, and how they are essential for our survival. We will also examine the complexities of the immune system, which plays a crucial role in protecting us from disease and infection.

Throughout this eBook, we will discuss the latest research and discoveries in the field of human anatomy and physiology. We will also look at how advancements in technology and medicine have helped us to better understand the human body and improve our ability to treat diseases and disorders.

Whether you are a student of biology, a healthcare professional, or simply someone with a curiosity about the human body, "The Wonders Within: Exploring the Human Body" is a must-read. It is a comprehensive guide that will leave you with a deeper appreciation for the amazing machine that is the human body. So, let's dive in and explore the wonders within!

Index

1. Understand the complex systems that make up the human body

2. Discover the function and structure of various organs in the body

3. Learn about the role of the immune system in protecting the body from disease

4. Explore the latest research and discoveries in the field of human anatomy and physiology

5. Gain a deeper appreciation for the amazing machine that is the human body

6. Understand the interconnectedness of the different systems in the body and how they work together

7. Learn how advancements in technology and medicine have helped us to better understand and treat the human body

8. Enhance your knowledge of biology and healthcare by reading this comprehensive guide

Chapter-1

Understand the complex systems that make up the human body

The human body is a complex system of various interdependent parts that work together to keep us alive and functioning. These systems range from the cardiovascular system that circulates blood throughout the body, to the nervous system that coordinates communication between various parts of the body. In this summary, we will explore the various systems that make up the human body and their functions.

The skeletal system provides structure and support for the body. It is made up of bones, cartilage, and ligaments. The skeletal system also serves as a storage site for minerals like calcium and phosphorus. Bones are connected to each other by joints, which allow for movement. There are three types of joints: immovable, slightly movable, and freely movable. The skull is an example of an immovable joint, while the elbow is an example of a freely movable joint.

The muscular system is responsible for movement and locomotion. It is made up of skeletal muscles, smooth muscles, and cardiac muscles. Skeletal muscles are attached to bones and are responsible for voluntary movement. Smooth muscles are found in organs like the stomach and intestines and are responsible for involuntary movement. Cardiac muscles make up the heart and are responsible for pumping blood throughout the body.

The circulatory system, also known as the cardiovascular system, is responsible for the circulation of blood throughout the body. It is made up of the heart, blood vessels, and blood. The heart is a muscular organ that pumps blood through the blood vessels. Blood vessels include arteries, veins, and capillaries. Arteries carry oxygen-rich blood away from the heart, while veins carry oxygen-poor blood back to the heart. Capillaries are tiny blood vessels that connect arteries and veins.

The respiratory system is responsible for breathing and the exchange of gasses. It is made up of the nose, mouth, pharynx, larynx, trachea, bronchi, and lungs. Air enters the body through the nose or mouth and travels down the pharynx and larynx before reaching the trachea. The trachea branches off into two bronchi, one for each lung. The bronchi continue to branch into smaller tubes called bronchioles, which lead to tiny air sacs called alveoli. It is in the alveoli where oxygen and carbon dioxide are exchanged between the lungs and the blood.

The digestive system is responsible for the breakdown and absorption of food. It is made up of the mouth, esophagus, stomach, small intestine, large intestine, rectum, and anus. Food is broken down into smaller molecules through the process of digestion. Enzymes in the mouth, stomach, and small intestine help break down food. Nutrients from the food are absorbed in the small intestine and passed on to the bloodstream for distribution to the rest of the body. The waste products of digestion are eliminated through the rectum and anus.

The urinary system is responsible for the elimination of waste products from the body. It is made up of the kidneys, ureters, bladder, and urethra. The kidneys filter waste products from the blood and produce urine. The ureters transport urine from the kidneys to the bladder, where it is stored until it is eliminated through the urethra.

The nervous system is responsible for communication and coordination between various parts of the body. It is made up of the brain, spinal cord, and nerves. The brain is responsible for receiving and interpreting information from the senses and initiating responses. The spinal cord is responsible for transmitting signals between the brain and the rest of the body. Nerves are responsible for transmitting signals between different parts of the body.

The endocrine system is responsible for producing and regulating hormones. It is made up of glands like the pituitary gland, thyroid gland, and adrenal glands. Hormones are chemical messengers that regulate various bodily functions such as growth, metabolism, and reproduction. The endocrine system works closely with the nervous system to maintain homeostasis, which is the body's ability to maintain a stable internal environment.

The immune system is responsible for protecting the body from foreign invaders such as bacteria, viruses, and fungi. It is made up of various cells, tissues, and organs including white blood cells, lymph nodes, and the spleen. The immune system works by identifying and attacking foreign substances while leaving healthy cells intact. When the immune system is compromised, it can lead to various diseases and infections.

The integumentary system is responsible for protecting the body from external damage and regulating temperature. It is made up of the skin, hair, and nails. The skin serves as a barrier against external threats such as bacteria and UV radiation. It also helps regulate body temperature through sweating and shivering.

In conclusion, the human body is a complex system made up of various interdependent parts that work together to keep us alive and functioning. Each system has a unique function, and they work together to maintain homeostasis and keep the body healthy. Understanding these systems and how they work can help us better care for our bodies and prevent diseases and illnesses.

Chapter-2

Discover the function and structure of various organs in the body

The human body is made up of many different organs that serve unique functions to keep us alive and functioning. In this summary, we will explore the structure and function of some of the major organs in the body.

The heart is a muscular organ located in the chest that is responsible for pumping blood throughout the body. The heart has four chambers: the right atrium, the left atrium, the right ventricle, and the left ventricle. Blood enters the heart through the atria and is pumped out through the ventricles. The heart is composed of cardiac muscle, which is specialized for contraction and can beat rhythmically without external stimulation. The heart is also surrounded by a protective sac called the pericardium.

The lungs are a pair of spongy organs located in the chest that are responsible for the exchange of oxygen and carbon dioxide. The lungs are composed of millions of small air sacs called alveoli, which are surrounded by blood vessels. When we inhale, air enters the lungs and oxygen is exchanged for carbon dioxide. The oxygen is then carried by the blood to the rest of the body, while the carbon dioxide is eliminated from the body through exhalation.

The liver is the largest internal organ in the body and is located in the upper right portion of the abdomen. The liver performs many vital functions, including filtering toxins from the blood, producing bile to aid in digestion, and storing glycogen for energy. The liver is also involved in the metabolism of drugs and other substances.

The kidneys are a pair of bean-shaped organs located in the lower back that are responsible for filtering waste products from the blood and producing urine. The kidneys receive blood from the renal artery and filter it through a complex network of tubules and blood vessels. The waste products are then eliminated from the body in the form of urine, while the useful substances are reabsorbed into the blood.

The brain is the most complex organ in the body and is responsible for controlling all bodily functions. The brain is composed of billions of nerve cells called neurons, which are organized into different regions that control different functions. The brain is protected by the skull and is cushioned by cerebrospinal fluid. The brain is also connected to the rest of the body through the spinal cord.

The pancreas is a glandular organ located in the abdomen that is responsible for producing hormones such as insulin and glucagon, which regulate blood sugar levels. The pancreas also produces enzymes that aid in the digestion of carbohydrates, proteins, and fats. The enzymes are released into the small intestine, where they break down food into smaller molecules for absorption.

The stomach is a muscular sac located in the upper abdomen that is responsible for storing and digesting food. The stomach produces hydrochloric acid and enzymes that break down food into smaller molecules. The food is then passed on to the small intestine for further digestion and absorption.

The small intestine is a long, narrow tube located in the abdomen that is responsible for the majority of nutrient absorption. The small intestine is lined with small finger-like projections called villi, which increase the surface area for nutrient absorption. The small intestine also produces enzymes that aid in the digestion of carbohydrates, proteins, and fats.

The large intestine, also known as the colon, is a wider tube located in the abdomen that is responsible for absorbing water and electrolytes from undigested food. The large intestine also contains beneficial bacteria that help break down indigestible substances and produce vitamins such as vitamin K.

The skin is the largest organ in the body and serves as a protective barrier against external threats. The skin is composed of three layers: the epidermis, the dermis, and the subcutaneous tissue. The epidermis is the outermost layer and is responsible for the formation of the skin barrier and the production of melanin, which gives skin its color. The dermis is the middle layer and contains hair follicles, sweat glands, and nerve endings. The subcutaneous tissue is the innermost layer and contains fat and connective tissue.

The spleen is a small organ located in the upper left portion of the abdomen that is responsible for filtering blood and removing old or damaged red blood cells. The spleen also produces white blood cells and antibodies, which help fight infections.

The gallbladder is a small organ located under the liver that stores and releases bile, which aids in the digestion of fats. Bile is produced by the liver and is transported to the gallbladder for storage. When we eat a fatty meal, the gallbladder releases bile into the small intestine to help break down the fats.

The adrenal glands are two small glands located on top of the kidneys that produce hormones such as adrenaline and cortisol, which help regulate the body's response to stress. Adrenaline increases heart rate and blood pressure, while cortisol increases blood sugar levels and suppresses the immune system.

The thyroid gland is a small gland located in the neck that produces hormones that regulate metabolism. The thyroid gland also plays a role in the development and growth of the body, as well as the regulation of body temperature.

In conclusion, the human body is made up of many different organs, each with its unique structure and function. These organs work together to keep us alive and functioning, and any disruption to one organ can have a significant impact on the entire body. Understanding the structure and function of these organs can help us better care for our bodies and prevent diseases and illnesses.

Chapter-3

Learn about the role of the immune system in protecting the body from disease

The immune system is a complex network of cells, tissues, and organs that work together to protect the body from harmful pathogens such as bacteria, viruses, and fungi. In this summary, we will explore the role of the immune system in protecting the body from disease.

The immune system has two main components: the innate immune system and the adaptive immune system. The innate immune system provides the body with immediate, nonspecific protection against pathogens, while the adaptive immune system provides more specific, targeted protection against specific pathogens.

The innate immune system includes physical barriers such as the skin and mucous membranes, as well as various cells such as neutrophils, macrophages, and natural killer cells. These cells work together to identify and destroy pathogens in the body.

The adaptive immune system, on the other hand, involves specialized cells called lymphocytes, including T cells and B cells, which are capable of recognizing and responding to specific pathogens. When a pathogen enters the body, it is recognized by lymphocytes, which then produce antibodies to attack and destroy the pathogen.

In addition to attacking and destroying pathogens, the immune system also plays a key role in the development of immunity. When the body is exposed to a pathogen for the first time, it may take several days for the immune system to produce enough antibodies to effectively fight the infection. However, once the immune system has been exposed to a pathogen, it can produce antibodies much more quickly and effectively in response to subsequent infections with the same pathogen. This process is known as immunity.

The immune system can also develop immunity through vaccination. Vaccines contain weakened or dead pathogens that stimulate the immune system to produce antibodies without causing disease. This allows the immune system to develop immunity to the pathogen without the risk of serious illness.

Despite its vital role in protecting the body, the immune system is not infallible. Some pathogens, such as HIV and tuberculosis, are able to evade the immune system and cause chronic infections. Additionally, the immune system can sometimes mistake healthy cells in the body for pathogens, leading to autoimmune disorders such as rheumatoid arthritis and lupus.

To support the immune system, it is important to maintain a healthy lifestyle. This includes eating a balanced diet rich in fruits and vegetables, exercising regularly, getting enough sleep, and avoiding smoking and excessive alcohol consumption. Additionally, it is important to practice good hygiene, such as washing hands regularly and covering the mouth and nose when coughing or sneezing, to prevent the spread of pathogens.

In some cases, the immune system may need additional support to fight off infections. This may involve the use of antibiotics to treat bacterial infections, antiviral medications to treat viral infections, or immunosuppressant drugs to treat autoimmune disorders. In some cases, immunotherapy may be used to boost the immune system's ability to fight cancer.

In conclusion, the immune system plays a critical role in protecting the body from disease. It is composed of a complex network of cells, tissues, and organs that work together to identify and destroy pathogens, develop immunity, and maintain overall health. While the immune system is not infallible, it can be supported through a healthy lifestyle and medical interventions when necessary. Understanding the immune system can help us better care for our bodies and prevent disease.

Chapter-4

Explore the latest research and discoveries in the field of human anatomy and physiology

The field of human anatomy and physiology is constantly evolving as new discoveries are made through research and technological advancements. In this summary, we will explore some of the latest research and discoveries in this field.

One area of research that has seen significant advancements is in the study of the human brain. Neuroimaging techniques such as magnetic resonance imaging (MRI) and positron emission tomography (PET) have allowed researchers to visualize the structure and activity of the brain in ways that were previously impossible. These techniques have led to new insights into the functions of different regions of the brain and how they interact with each other.

One notable discovery in the field of neuroscience is the concept of neuroplasticity. This refers to the brain's ability to reorganize and form new neural connections in response to changes in the environment or experiences. This has important implications for the treatment of brain injuries and neurological disorders, as it suggests that the brain may be able to adapt and recover function even after significant damage.

Another area of research that has seen significant advancements is in the study of the human microbiome. The microbiome refers to the collection of microorganisms, including bacteria, viruses, and fungi, that live in and on the human body. Recent research has shown that the microbiome plays a critical role in maintaining overall health, including regulating the immune system, aiding in digestion, and even influencing mood and behavior.

One notable discovery in this field is the association between changes in the gut microbiome and the development of certain diseases, such as inflammatory bowel disease and obesity. This has led to the development of new treatments and therapies that aim to manipulate the microbiome to improve health outcomes.

Advancements in technology have also led to new discoveries in the field of genetics. The development of high-throughput sequencing techniques has allowed researchers to analyze large amounts of genetic data quickly and efficiently, leading to new insights into the genetic basis of diseases and conditions.

One notable discovery in this field is the identification of genetic variants associated with an increased risk of developing certain types of cancer. This has led to new approaches to cancer screening and prevention, including the development of targeted therapies that are tailored to specific genetic mutations.

Finally, research in the field of anatomy has led to new discoveries in our understanding of the structure and function of the human body. For example, recent research has shown that the lymphatic system, which was previously thought to only exist in certain parts of the body, is actually much more extensive and plays a critical role in the immune system.

Another area of research that has seen significant advancements is in the study of biomechanics, or the application of mechanical principles to biological systems. This has led to new insights into the mechanics of movement, including the development of new prosthetics and other assistive technologies that can help people with disabilities or injuries regain mobility.

Overall, the latest research and discoveries in the field of human anatomy and physiology have led to new insights into the structure and function of the human body, as well as new approaches to the diagnosis, treatment, and prevention of diseases and conditions. As technology continues to evolve and new research is conducted, it is likely that we will continue to uncover new information about the complex systems that make up the human body and how they interact with each other.

Chapter-5

Gain a deeper appreciation for the amazing machine that is the human body

The human body is a remarkable machine that is capable of a vast range of functions and abilities. From the intricacies of our cells and organs to the complex systems that govern our movements, thoughts, and emotions, the human body is a marvel of biological engineering that deserves our awe and admiration.

At the most fundamental level, the human body is made up of trillions of cells, each with its own unique role and function. These cells are organized into tissues, which in turn make up organs such as the heart, lungs, brain, and liver. Together, these organs form systems that work together to keep the body functioning properly.

One of the most important systems in the human body is the nervous system, which is responsible for controlling and coordinating all of the body's movements and functions. The nervous system consists of the brain, spinal cord, and a network of nerves that extend throughout the body. Through this system, we are able to perceive the world around us, control our movements, and communicate with others.

Another essential system is the circulatory system, which is responsible for transporting oxygen, nutrients, and other essential substances throughout the body. This system includes the heart, blood vessels, and blood. The heart is an incredibly powerful muscle that pumps blood throughout the body, while the blood vessels act as a network of highways, carrying blood to every part of the body.

The respiratory system is also crucial to our survival, as it allows us to breathe in oxygen and exhale carbon dioxide. This system includes the lungs, trachea, bronchi, and other structures that work together to facilitate the exchange of gasses in the body.

Our digestive system is responsible for breaking down food and extracting nutrients from it, which are then used by the body for energy and growth. This system includes the mouth, esophagus, stomach, intestines, and other organs that work together to process food and eliminate waste.

In addition to these systems, the human body also has a number of other complex structures and functions. For example, our immune system is responsible for protecting us from harmful pathogens and other invaders, while our endocrine system regulates hormones and other chemicals that affect our growth, metabolism, and other bodily functions.

Our ability to move and interact with the world around us is also a testament to the incredible complexity and sophistication of the human body. Our muscles, bones, and joints work together in a finely-tuned system that allows us to walk, run, jump, and perform a wide range of other movements.

At the same time, our brain is constantly processing information from our senses, allowing us to perceive the world around us and make decisions based on that information. Our ability to think, reason, and solve problems is a hallmark of human intelligence, and it is made possible by the complex interplay of neurons and other cells in the brain.

Finally, the human body also has an incredible capacity for adaptation and self-healing. When we get injured or sick, our body is able to mount a response to repair the damage and fight off the infection. Even as we age, our body is able to adapt to changing conditions and maintain a level of functioning that allows us to continue living and thriving.

In conclusion, the human body is a truly amazing machine that deserves our admiration and respect. From the tiniest cells to the most complex systems, every part of the body is intricately interconnected and finely-tuned to perform its role in keeping us alive and healthy. Whether we are running a marathon, solving a complex math problem, or simply enjoying a beautiful sunset, it is the incredible machine that is the human body that makes it all possible.

Chapter-6

Understand the interconnectedness of the different systems in the body and how they work together

The human body is made up of various interconnected systems, each with a specific function and purpose. These systems work together in a complex web of interactions to keep the body functioning properly and maintain overall health.

The nervous system is one of the most important systems in the body, as it controls and coordinates all other systems. It is made up of the brain, spinal cord, and network of nerves that extend throughout the body. The nervous system receives input from the environment through the senses and processes this information to initiate appropriate responses in the body. It also controls all voluntary and involuntary movements, such as the beating of the heart, the movement of food through the digestive system, and the dilation of blood vessels.

The endocrine system is another important system that works closely with the nervous system. It is responsible for the production and secretion of hormones, which regulate various bodily functions, such as metabolism, growth, and sexual development. The endocrine system includes glands such as the pituitary, thyroid, and adrenal glands, as well as the pancreas and ovaries or testes.

The circulatory system is responsible for the transport of oxygen, nutrients, and other essential substances throughout the body. It includes the heart, blood vessels, and blood. The heart pumps blood through the arteries and veins, while the blood delivers oxygen and nutrients to the cells and removes waste products. The lymphatic system, which works closely with the circulatory system, is responsible for removing excess fluids and waste from tissues and helping to fight infections.

The respiratory system is responsible for the exchange of gasses between the body and the environment. It includes the lungs, trachea, bronchi, and other structures that allow air to flow into and out of the body. Oxygen is taken up by the blood and delivered to the cells, while carbon dioxide, a waste product of cellular metabolism, is removed from the body through the lungs.

The digestive system is responsible for the breakdown and absorption of food and nutrients. It includes the mouth, esophagus, stomach, intestines, and other organs that work together to digest food and eliminate waste products. The liver and pancreas also play important roles in the digestive system, producing digestive enzymes and regulating blood sugar levels.

The urinary system is responsible for the elimination of waste products from the body. It includes the kidneys, ureters, bladder, and urethra. The kidneys filter waste products from the blood and produce urine, which is then excreted from the body through the ureters, bladder, and urethra.

The musculoskeletal system is responsible for movement and support of the body. It includes bones, muscles, tendons, and ligaments. Bones provide the framework for the body, while muscles allow movement and provide strength and support. Tendons connect muscles to bones, while ligaments connect bones to other bones.

Finally, the immune system is responsible for defending the body against foreign invaders, such as bacteria, viruses, and parasites. It includes a complex network of cells and tissues that work together to identify and destroy harmful pathogens. The immune system is also responsible for recognizing and eliminating abnormal cells, such as cancer cells.

All of these systems are interconnected and work together to maintain overall health and functioning of the body. For example, the circulatory system provides oxygen and nutrients to the cells, while the respiratory system ensures that the body has a steady supply of oxygen. The digestive system breaks down food into nutrients that are absorbed by the circulatory system, which then delivers these nutrients to the cells.

The nervous system and endocrine system work together to regulate various bodily functions, such as heart rate, blood pressure, and digestion. The urinary system helps to maintain the body's balance of fluids and electrolytes, while the immune system helps to protect against infections and diseases.

In conclusion, the human body is an amazing machine that is capable of incredible feats. The interconnectedness of its different systems is what makes it so complex and remarkable. Each system relies on the others to function properly, and any disruption in one system can have a cascading effect on the others.

Understanding how these systems work together can help us appreciate the complexity of the human body and the importance of maintaining good health. Proper nutrition, exercise, and medical care are all essential for keeping these systems functioning at their best.

Despite our advanced understanding of the human body, there is still much to learn about how it works and how we can better care for it. Ongoing research and technological advancements will continue to shed light on the intricacies of the human body, helping us to further appreciate the amazing machine that it is.

Chapter-7

Learn how advancements in technology and medicine have helped us to better understand and treat the human body

Advancements in technology and medicine have revolutionized our understanding of the human body and how we can treat it. These innovations have enabled us to diagnose and treat diseases more effectively, develop new therapies, and improve our overall health and well-being.

One of the most significant technological advancements in medicine has been the development of medical imaging techniques. X-rays, CT scans, MRI scans, and other imaging technologies allow doctors to visualize the inside of the body without invasive procedures. These techniques can help to diagnose a wide range of medical conditions, from broken bones to cancer. They also allow doctors to monitor the progression of diseases and the effectiveness of treatments.

Another area of medical technology that has seen rapid advancement is genomics. Genomics is the study of the genetic material of an organism. With the advent of new DNA sequencing technologies, researchers can now analyze large amounts of genetic data more quickly and accurately than ever before. This has led to the discovery of many new genes and genetic mutations associated with various diseases.

Advances in genomics have also enabled the development of personalized medicine. Personalized medicine uses genetic information to tailor medical treatments to an individual's unique genetic makeup. This approach allows doctors to prescribe medications and treatments that are more likely to be effective and less likely to cause adverse side effects.

Nanotechnology is another field that has seen significant advances in recent years. Nanotechnology involves the manipulation of materials on a molecular or atomic scale. In medicine, nanotechnology is being used to develop new drugs and drug delivery systems that can target specific cells or tissues in the body. These nanomedicines are more precise and efficient than traditional drug delivery methods, which can result in less damage to healthy tissues and fewer side effects.

Advances in robotics and artificial intelligence are also transforming medicine. Robotic surgery allows doctors to perform complex surgical procedures with greater precision and control. Artificial intelligence is being used to analyze large amounts of medical data and identify patterns that can help to diagnose diseases and develop new treatments.

Telemedicine is another area that has seen rapid growth in recent years. Telemedicine involves the use of telecommunications technology to provide medical care and advice remotely. This approach has become increasingly popular during the COVID-19 pandemic, as it allows doctors to consult with patients without putting them at risk of infection. Telemedicine has also made medical care more accessible to people who live in remote or underserved areas.

The development of new drugs and therapies is another area where technology and medicine are advancing rapidly. Traditional drug discovery methods can take years or even decades to produce new treatments. However, with the help of computer simulations and artificial intelligence, researchers can now design and test new drugs more quickly and efficiently.

One promising area of research is immunotherapy. Immunotherapy involves using the body's own immune system to fight cancer and other diseases. This approach has shown promising results in treating a variety of cancers, including melanoma, lung cancer, and bladder cancer.

Advancements in technology and medicine have also led to a better understanding of the human body and its functioning. For example, new imaging technologies have allowed researchers to study the brain in more detail, leading to a better understanding of how it works and how it can be affected by diseases such as Alzheimer's and Parkinson's.

In conclusion, advancements in technology and medicine have had a profound impact on our understanding of the human body and how we can treat it. These innovations have enabled us to diagnose and treat diseases more effectively, develop new therapies, and improve our overall health and well-being. As technology continues to advance, we can expect to see even more breakthroughs in medicine and an ever-improving understanding of the amazing machine that is the human body.

Chapter-8

Enhance your knowledge of biology and healthcare by reading this comprehensive guide

Biology and healthcare are complex and fascinating fields that are constantly evolving with new discoveries and advancements. To enhance your knowledge of these fields, it is essential to have a comprehensive understanding of the fundamental concepts, principles, and practices that underlie them.

At its core, biology is the study of life and living organisms. It encompasses everything from the structure and function of cells to the interactions between organisms and their environment. Understanding biology is essential for understanding the causes and mechanisms of diseases, as well as developing new treatments and therapies.

Healthcare, on the other hand, is the field of medicine that focuses on the prevention, diagnosis, and treatment of diseases and injuries. Healthcare professionals work to promote good health and well-being by providing medical care, advice, and education to patients and the public.

One of the key concepts in biology and healthcare is genetics. Genetics is the study of genes and heredity, and how they influence the traits and characteristics of organisms. Advances in genetics have enabled researchers to identify genetic mutations and variations that are associated with diseases such as cancer, Alzheimer's, and diabetes.

Another important concept in biology and healthcare is cell biology. Cells are the basic unit of life, and understanding their structure and function is essential for understanding the workings of the body. Advances in cell biology have led to the development of new therapies, such as stem cell therapy, which uses stem cells to repair damaged tissues and organs.

Physiology is another key concept in biology and healthcare. Physiology is the study of the functions of living organisms and their parts. Understanding physiology is essential for understanding how the body works and how it can be affected by disease or injury.

Anatomy is also an important concept in biology and healthcare. Anatomy is the study of the structure of living organisms and their parts. Understanding anatomy is essential for understanding the relationships between different parts of the body and how they function together.

Medical imaging is a crucial tool in healthcare that allows doctors to visualize the inside of the body without invasive procedures. There are several types of medical imaging technologies, including X-rays, CT scans, MRI scans, and ultrasound. These technologies are used to diagnose a wide range of medical conditions, from broken bones to cancer.

Another important aspect of healthcare is disease prevention. Preventive medicine involves measures that are taken to prevent the occurrence of diseases or injuries. This can include vaccinations, screening tests, and lifestyle changes such as exercise and healthy eating.

Treatment and management of diseases and injuries are also critical components of healthcare. Treatment can include medications, surgeries, therapies, and other interventions. Healthcare professionals work to develop personalized treatment plans that are tailored to each individual patient's unique needs and circumstances.

Medical research is a crucial aspect of healthcare that is essential for advancing our understanding of diseases and developing new treatments and therapies. There are many different types of medical research, including basic research, clinical research, and translational research. Each type of research plays a crucial role in advancing medical knowledge and improving patient care.

Public health is another important aspect of healthcare that focuses on the health of populations rather than individuals. Public health professionals work to promote healthy lifestyles, prevent the spread of diseases, and improve access to healthcare services.

In conclusion, biology and healthcare are complex and fascinating fields that are essential for understanding the workings of the human body and improving our health and well-being. Understanding the fundamental concepts, principles, and practices that underlie these fields is essential for anyone interested in pursuing a career in healthcare or biology. By staying up-to-date with the latest advancements and discoveries, we can continue to improve patient care, develop new therapies, and enhance our understanding of the amazing machine that is the human body.

www.ingramcontent.com/pod-product-compliance
Lightning Source LLC
Chambersburg PA
CBHW041949240526
45473CB00036B/2797